Scientists Who Changed the World

Albert Einstein

Anita Croy

CRABTREE
PUBLISHING COMPANY
WWW.CRABTREEBOOKS.COM

CRABTREE
PUBLISHING COMPANY
WWW.CRABTREEBOOKS.COM

Author: Anita Croy
Editors: Sarah Eason, Melissa Boyd, Ellen Rodger
Proofreader and indexer: Jennifer Sanderson
Proofreader: Wendy Scavuzzo
Editorial director: Kathy Middleton
Design: Paul Myerscough and Lynne Lennon
Photo research: Rachel Blount
Print coordinator: Katherine Berti

Written, developed, and produced by Calcium

Photo Credits:
t=Top, c=Center, b=Bottom, l= Left, r=Right

Inside: Library of Congress: Al. Aumuller: p. 46; NASA: JPL-Caltech: p. 52; Shutterstock: Afanasiev Andrii: p. 35; Billion Photos: p. 27; BlueRingMedia: p. 25; Marusya Chaika: p. 7b; Designua: p. 30; EQRoy: p. 49; Everett Historical: pp. 38, 41, 45; Everst: p. 55; F11photo: p. 9; David Fowler: p. 60; IgorZh: p. 28; Jasmine_K: p. 12; LaMiaFotografia: p. 7c; Lario: p. 31; MarkoV87: p. 22; Morphart Creation: p. 20; Natali Art Collections: p. 37; Nerthuz: p. 15; Nicku: p. 14; Vadim Petrakov: p. 34; Galyamin Sergej: p. 26; Sumire8: p. 59t; Vchal: pp. 18, 29; Ken Wolter: p. 17; Wikimedia Commons: pp. 6, 10, 11, 19, 39, 40, 59c; Lucien Chavan: p. 24; Maud Cuney-Hare: p. 47; Energy.gov: p. 44; ESA/Hubble & NASA: p. 50; Adam Evans: p. 32; Max Halberstadt: p. 51; Anders Wilhelm Karnell: p. 58; NASA: p. 16; NASA, ESA, the Hubble Heritage Team (STScI/AURA), A. Nota (ESA/STScI), and the Westerlund 2 Science Team: p. 61; National Photo Company Collection: p. 36; Wilhelm Conrad Röntgen/Wellcome Images: p. 21; Roxaneweb: p. 57b; Ruffnax: p. 54; Ferdinand Schmutzer: p. 4; Lucas Taylor/CERN: p. 56; Tighef: p 57t; Orren Jack Turner, Princeton, N.J.: p. 42; United Press/Library of Congress: p. 48; Gottfried Wolfsgruber: p. 8.

Library and Archives Canada Cataloguing in Publication

Title: Albert Einstein / Anita Croy.
Names: Croy, Anita, author.
Description: Series statement: Scientists who changed the world | Includes index.
Identifiers: Canadiana (print) 20200225723 | Canadiana (ebook) 20200225731 | ISBN 9780778782094 (hardcover) | ISBN 9780778782230 (softcover) | ISBN 9781427126092 (HTML)
Subjects: LCSH: Einstein, Albert, 1879-1955—Juvenile literature. | LCSH: Physicists—Biography—Juvenile literature. | LCGFT: Biographies.
Classification: LCC QC16.E5 C76 2021 | DDC j530.092—dc23

Library of Congress Cataloging-in-Publication Data

Names: Croy, Anita, author.
Title: Albert Einstein / Anita Croy.
Description: New York : Crabtree Publishing Company, [2021] | Series: Scientists who changed the world | Includes index.
Identifiers: LCCN 2020017179 (print) | LCCN 2020017180 (ebook) | ISBN 9780778782094 (hardcover) | ISBN 9780778782230 (paperback) | ISBN 9781427126092 (ebook)
Subjects: LCSH: Einstein, Albert, 1879-1955--Juvenile literature. | Physicists--Biography--Juvenile literature.
Classification: LCC QC16.E5 C79 2021 (print) | LCC QC16.E5 (ebook) | DDC 530.092 [B]--dc23
LC record available at https://lccn.loc.gov/2020017179
LC ebook record available at https://lccn.loc.gov/2020017180

Crabtree Publishing Company
www.crabtreebooks.com 1-800-387-7650

Printed in the U.S.A./082020/CG20200601

Published in Canada
Crabtree Publishing
616 Welland Ave.
St. Catharines, Ontario
L2M 5V6

Published in the United States
Crabtree Publishing
347 Fifth Ave
Suite 1402-145
New York, NY 10016

Published in the United Kingdom
Crabtree Publishing
Maritime House
Basin Road North, Hove
BN41 1WR

Published in Australia
Crabtree Publishing
3 Charles Street
Coburg North
VIC, 3058

Contents

CHAPTER 1

This photograph shows Albert Einstein in 1921, the year he won the **Nobel Prize** for **Physics**.

ALBERT EINSTEIN
Biography

Born: March 14, 1879

Place of birth: Ulm, Germany

Mother: Pauline Koch Einstein

Father: Hermann Einstein

Famous for: The **equation** $E=mc^2$, states that energy, or the power to do work, and mass (weight) are the same thing in different forms. This is called the world's most famous equation.

How he changed the world: Einstein's **Theory of Relativity** changed our understanding of space and launched the new science of cosmology, or the study of the universe and its birth and development. Many of his ideas, such as the existence of **black holes**, could not be proven until after his death, when technology had improved.

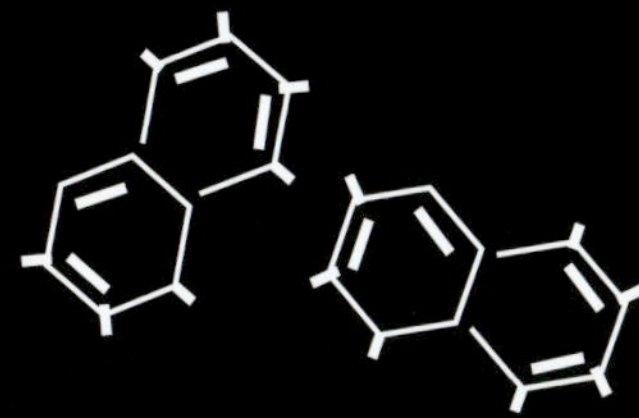

Einstein said
that the IMPORTANT
thing is to not
stop QUESTIONING.
Curiosity has its
own REASON
for EXISTING.

MATH BOY

Albert Einstein was born into a Jewish family in the town of Ulm, Germany, in 1879. A year later, the family moved to the large city of Munich. Einstein's father was a businessman, while his mother stayed at home and raised Albert and his younger sister, Maria, who was known as Maja.

This photo of Albert and Maja was taken in about 1886, when Albert was seven.

When Albert was five years old, he saw a compass for the first time. It fascinated him. He wanted to know about the invisible forces that moved the needle. Despite his curiosity about how things worked, Albert was a poor student at school. He hated the rigid structure of lessons. He was teased and even attacked by his classmates for being a Jew.

Albert was a poor student at school.

Most of Albert's learning took place outside of school, thanks to a medical student named Max Talmey, who ate with the Einsteins every Thursday night. When Albert was 12, Max gave him a book about the branch of mathematics called geometry. The book, by the ancient Greek mathematician Euclid, was called *Elements*. Albert called it his "sacred little geometry book." Max became Albert's unofficial tutor. He also introduced Albert to science and **philosophy**. Albert worked on his geometry problems all week and showed them to Max every Thursday. Before long, the student had overtaken his teacher.

Ideas that changed the world

Einstein thought that an education based on freedom of choice and being responsible for one's own learning was far superior to an education that relied on rules and discipline, obeying authority without question, and ambition.

Exploring the ideas

Like English scientist Sir Isaac Newton (1642–1727), whose theories he would later challenge, Einstein did not enjoy school. He did most of his learning outside of school. When Einstein was 15, his family moved to Italy for his father's new business, but Einstein stayed in Germany to finish school. He hated the way the German school system forced students to learn without questioning. Einstein spent his whole life questioning and challenging accepted ideas, then coming up with ideas of his own.

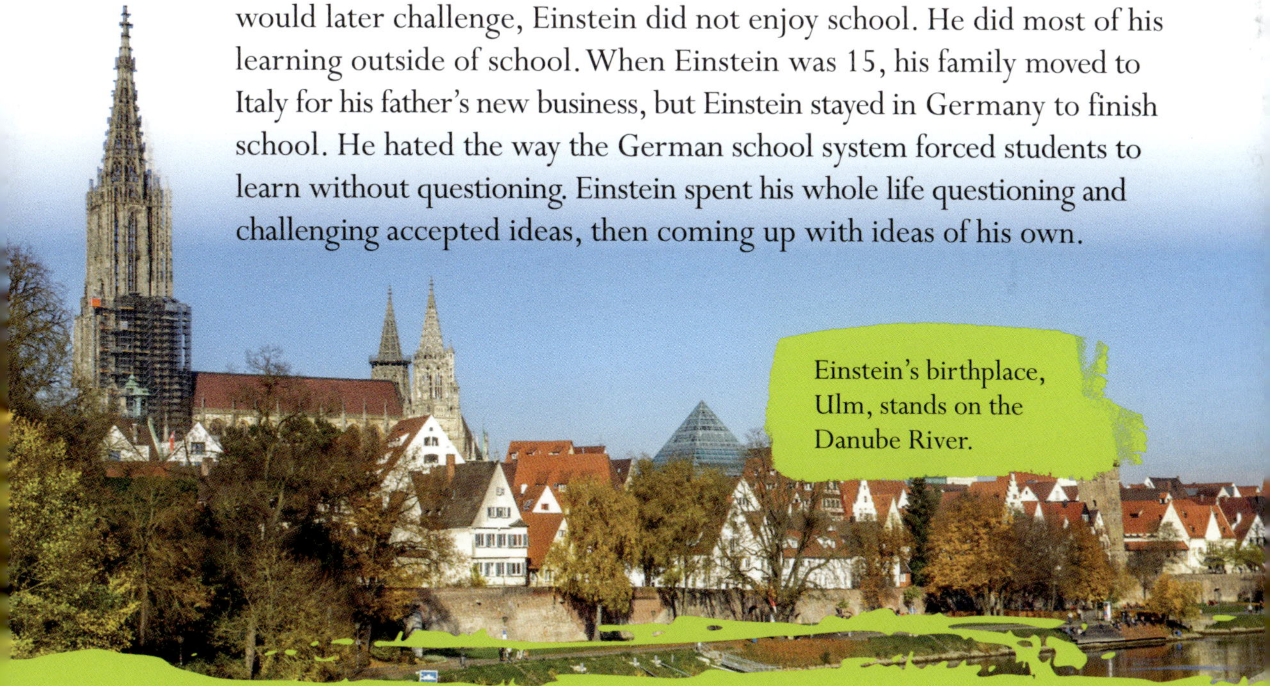

Einstein's birthplace, Ulm, stands on the Danube River.

HISTORY'S STORY

The Einsteins were Jewish, but did not attend synagogue regularly. They adapted Jewish customs to suit their lives, such as the Friday night Sabbath meal. Instead of inviting Max Talmey over on Fridays, as was the Jewish custom, they invited him on Thursdays. Einstein was picked on at school for being Jewish. This might have encouraged him to stand up for the underdog, or people not expected to succeed, throughout his life.

A FRESH START

When he was 15, Albert dropped out of school in Munich and joined his family in Milan, Italy. With no school to attend, he helped out in his father's factory and hung out in museums and libraries. He soon decided he wanted to study at the Polytechnic Institute in Zürich, Switzerland. The only problem was that the "Poly" had a tough entrance examination. It also had an age requirement of 18.

At age 16, Einstein's physics and math abilities were already far ahead of those of his classmates.

Einstein did so well on his examination that the Poly agreed to take him, but only if he got his high-school diploma first. Albert chose a school in Aarau, Switzerland, that encouraged independent thinking. He was already asking himself questions such as: What would it be like to travel on a beam of light? Could a light **wave** stand still? These were not questions covered in the classroom.

In Aarau, Einstein graduated second in his class in June 1896. He reapplied to the Poly and was accepted to study math and physics. The Poly suited Einstein perfectly. In four years of study, students only took two sets of exams. There was no homework, and students did not have to attend lectures if they chose not to.

Zürich, the largest city in Switzerland, has a beautiful setting on a lake.

Falling in love

While at the Poly, Albert skipped lectures and spent his time reading. When it was time for tests, he borrowed notes from his friend Marcel Grossmann. Albert started to date the only woman among the other four physics students. Mileva Maric came from Titel in present-day Serbia, which was then part of Austria-Hungary. Women were not allowed to attend college there, which was why Mileva had come to Zürich to study.

Once Mileva and Albert were a couple, his attendance at the Poly fell still further, much to the anger of his professors. One complained that, even though Einstein was smart, "You'll never let yourself be told anything." Einstein ignored his professors. In the final exam, he came fourth while Mileva came fifth in the class. At around this time, Albert gave up his German nationality to avoid being forced to join the army. He applied to be a Swiss citizen. It took five years before he was granted Swiss nationality in 1901.

Losing a child

Einstein's family were not pleased with his relationship with Mileva, and were even less pleased when the couple had a daughter, Lieserl, who was born in 1902 before they were married. They kept the birth secret from almost everyone. Nobody knows for sure what happened to the baby. Historians think she may have died from a disease called scarlet fever.

A PERFECT JOB

After he graduated, Einstein could not get a job. Without good reports from his professors, he could not get the university teaching post he really wanted. He had to support Mileva, so he was desperate for money. His father's latest business venture had failed, so Einstein could not rely on his family for money. He did odd jobs teaching math and physics, but managed to get fired from one position for arguing with his boss.

Einstein and Mileva had three children, but only two survived to adulthood.

Einstein's friend Marcel Grossmann learned there was an opening at the Swiss **Patent** Office in Bern, Switzerland. People applied to this office when they had invented something new, hoping to be given a patent. This is a government license recognizing that the invention belongs entirely to that person. Grossmann's father was a friend of Friedrich Haller, who ran the office. On the advice of Grossmann's father, Haller interviewed Einstein for two hours. Then came an agonizingly long wait to find out if he had the job.

Hoping the job would be his, Einstein moved to Bern with Mileva in late 1901. It was not until June 23, 1902, that he finally started work. His job was to help examine new inventions. When he looked at a new invention, Einstein was told by Haller to assume everything was wrong. Einstein was the perfect person for the job because he questioned everything.

Making time for time

Even though he worked six days a week, Einstein still found time to continue with his own research. One of the patents he worked on was about new electrical technology to coordinate the Swiss railroad system. Bern used its famous old central clock tower to set all the city's clocks, including those at the train station. Would it be possible to copy a similar scheme across the whole country?

Einstein thought about the physics involved. He knew that electricity was not instantaneous, as electrical current moved at the speed of light. He wondered whether every clock could be set to exactly the same time, and he also began to consider whether it was possible for two events to happen at exactly the same moment. His answer to these questions would become key to his **Special Theory of Relativity**.

Bern's old clock tower was a tourist attraction in Einstein's day. Visitors still come to watch mechanical figures move around on the hour.

HISTORY'S STORY

Einstein married Mileva on January 6, 1903. They had two sons, born in 1904 and 1910. For several years, the family lived happily together, but when Albert left for Berlin in 1914, Mileva and their sons stayed in Zürich. The couple divorced in 1919. Later that year, Albert married his distant cousin, Elsa, and remained happily married to her until her death in 1936.

CHAPTER 2

When Einstein began his career, people thought the Milky Way **galaxy** (shown here as a pale arch) was the entire universe.

A CHANGING WORLD
Background

Einstein worked at a time when scientists were using new theories and techniques to reconsider some of the most basic questions about how the universe worked. Some major areas of research included:

a. The nature of light, including whether it travels as waves or tiny **particles**

b. Motion in the universe and the laws that govern it

c. The nature of space, including how it relates to time

d. The structure of atoms, which are the tiniest parts of any substance that can exist alone

e. The relationship between electrical forces and moving objects, known as **electrodynamics**

Einstein said he was in awe when he thought about the MYSTERIES of ETERNITY, of LIFE, of the marvelous structure of REALITY. He said it is enough for a scientist merely to understand a LITTLE of this mystery each day.

NEWTON'S LAWS

When Einstein began his career, ideas about the universe were still governed by the laws figured out by the English physicist and mathematician Sir Isaac Newton. Newton suggested that the universe could be explained by mathematical equations. His three Laws of Motion seemed to explain everything, from why an apple falls from a tree to why the planets move around the Sun.

Einstein's work would replace Newton's theories.

Newton came up with his laws in 1666, when he was only 24 years old. Newton had returned home to the English countryside from Cambridge University, where he was a student. The university had closed to stop the spread of a deadly disease called the plague. Newton had time to sit and think. During what is called his *annus mirabilis*, or miracle year, he came up with his three Laws of Motion and also explained that everything on Earth was pulled downward by the planet's **gravity**. Newton also studied the nature of light. Most of Newton's laws are still used today. However, Einstein's work replaced Newton's theories for **astronomers** looking at stars and **physicists** studying the atom.

Sir Isaac Newton said that he was inspired to come up with the **theory** of gravity when he saw an apple fall from a tree.

Ideas that changed the world

Einstein said no one should think that his Theory of Relativity or any other theory could overthrow Newton's great creations in any real sense. He said that Newton's clear and wide ideas would forever keep their importance as the foundation on which modern physics was built.

Exploring the ideas

Einstein, like all physicists, was very grateful to Sir Isaac Newton. Newton himself had acknowledged he was grateful to the scientists who came before him, such as Nicolaus Copernicus (1473–1543), Galileo Galilei (1564–1642), and Johannes Kepler (1571–1630). They had all argued that Earth revolved around the Sun, rather than the other way around as the ancient Greeks had suggested. Newton's theories formed the basis of modern physics. Later physicists did not overturn his ideas, but added to them with new theories as knowledge of the universe grew.

The Sun holds the planets in orbit with its massive gravity.

HISTORY'S STORY

The 1800s were a time of great change in **astronomy**. Astronomers went from simply cataloging, or listing, objects in the night sky to using advances in math, **chemistry**, physics, and **geology** to figure out the nature of the parts of space they could not see. Using mathematical equations, astronomers discovered the existence of the planet Neptune in 1846.

SHAPE OF THE UNIVERSE

Newton had explained how gravity held the universe together, by pulling all objects toward each other. However, there were many questions about space that remained unanswered. There were even more questions that had yet to be thought of. Early in the 1900s, astronomers were just beginning to wonder how big space was and how it began.

To answer the questions astronomers were now asking, they had to have the right tools to work with. The first thing they needed to do was to allow their minds to expand. Improved telescopes had allowed people to see much deeper into space. It was becoming clear that the solar system and the rest of the universe were much, much bigger than anyone could imagine.

Newton's theory of gravity predicted the existence of the planet Neptune, which was discovered in 1846.

At the start of the 1900s, astronomers did not know that the universe was expanding. They shared the beliefs of earlier astronomers that the universe was fixed. Most believed that the Sun, Earth and the other planets, and the billions of stars of the Milky Way galaxy made up the entire universe. As late as 1905, the idea that an infinite universe lay beyond the Milky Way was no more than a mathematical possibility.

More galaxies

Around the time Einstein came up with a remarkable series of physics breakthroughs in 1905, the modern idea of the universe started to emerge thanks to powerful new telescopes. These showed there were billions of galaxies beyond the Milky Way. Astronomers recognized that the fuzzy lights they could see were galaxies millions of miles from our own. In 1923, the American astronomer Edwin Hubble was able to calculate the distance of the Andromeda galaxy from Earth. Hubble noticed what he thought were exploding stars, but then realized that they were growing brighter and dimming in a regular pattern. He used these stars, now called Cepheid variable stars, and some complicated math to figure out that Andromeda was 900,000 light-years away (a light-year is the distance light can travel in one year, or about 6 trillion miles (9.7 trillion km)).

Mysterious space

The more astronomers looked into the sky, the more they realized how much about it remained a mystery. They did not know what space was made up of and what, if anything, lay in the vast distances between the stars. Theories were being suggested of a **Big Bang** that brought the universe into existence but nobody had the math to prove how that might have happened. That theory would not be confirmed until later in the century.

Edwin Hubble discovered galaxies using the Mount Wilson Observatory in California, which opened in 1917.

A NEW PHYSICS

Newton's Laws of Motion explained the movements of the solar system, as well as the behavior of atoms. Newton had first suggested that atoms were not fixed but were "swimming" everywhere. More than 200 years later, physicists started to investigate a world of particles much smaller than atoms, existing inside atoms. These are called subatomic particles.

... a new set of laws was needed to describe the universe.

In the 1500s and 1600s, Galileo and Newton came up with theories that transformed physics and astronomy. They carried out thousands of experiments to prove their ideas. By the start of the 1900s, many physicists were interested in studying subjects that could not be observed. This required a new branch of physics called **theoretical physics**, which uses math and models to explore and prove theories about things that cannot actually be seen. They proved theories by solving problems on paper rather than by carrying out experiments. This new physics recognized that a new set of laws was needed to describe the universe. More than 20 percent of the **mass** they believed must be contained in galaxies could not be seen and could not be explained using existing laws.

Subatomic particles called **protons** and **neutrons** form the heart of the atom, while **electrons** whiz around them—but most of an atom is empty space.

Two new theories

By Einstein's time, scientists had realized that Newton's laws could not explain the behavior of subatomic particles. A new method, called **quantum mechanics**, was created to deal with them. Scientists knew that understanding subatomic particles would help explain how the universe came about.

Today, scientists use two modern theories in addition to Newton's laws to describe the universe. One is Einstein's General Theory of Relativity and the other is quantum mechanics. Relativity describes the large-scale structure of the universe. Quantum mechanics deals with its subatomic particles, which might be as tiny as a millionth of a millionth of an inch. However, scientists soon realized that the two theories do not work together. Today's scientists are searching for a quantum theory that will explain gravity—but that theory has not been thoroughly **formulated** yet.

The electrons discovered by J.J. Thomson are the reason electricity flows as energy.

HISTORY'S STORY

An atom is the smallest particle of anything that can exist. It is made up of at least three subatomic particles: electrons, neutrons, and protons. In 1897, the English physicist J.J. Thomson discovered the first subatomic particle: the electron. Today, scientists believe there are even smaller subatomic particles and are trying to identify them.

NINETEENTH-CENTURY BREAKTHROUGHS

1801 Wave theory of light introduced
Thomas Young suggests that light travels in invisible waves.

1803 Atomic theory of matter introduced
John Dalton suggests that **matter** is made up of tiny atoms.

1820 Electromagnetism discovered
The Danish scientist Hans Christian Ørsted realizes that electricity and magnetic fields have unusual effects where they interact.

1824 Carnot cycle described
Nicolas Léonard Sadi Carnot describes the expansion and contraction of gases in a heat engine, a machine designed to change heat energy into mechanical energy.

1831 Principles of electromagnetic induction discovered
Michael Faraday uses moving magnetic fields to create a flow of electrical current.

Michael Faraday used magnets to create electricity in a closely wound spool of wire.

1843 First law of thermodynamics introduced
The Frenchman James Prescott Joule argues that heat is a form of energy, so it cannot be created or destroyed.

1847 Law of conservation of energy introduced
Hermann von Helmholtz builds on Joule's idea to argue that the energy in a closed system remains constant.

1859 Kinetic theory developed
James Clerk Maxwell argues that the properties of a material are created by the motions of its atoms and **molecules**.

1865 Theory of electromagnetism developed
James Clerk Maxwell shows that electric and magnetic fields travel through space as waves at the speed of light. Light is an electromagnetic field.

1895 X-rays discovered
The German scientist Wilhelm Conrad Röntgen uses a form of invisible radiomagnetic waves to take images of the insides of people's bodies.

1896 Radioactivity discovered
The French physicist Henri Becquerel realizes that some atoms give off radiation, or damaging, high-energy particles.

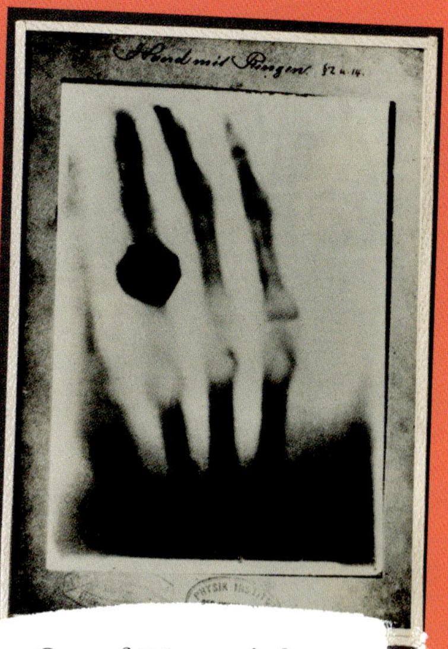

One of Röntgen's first X-rays showed his wife's hand—wearing a ring.

CHAPTER 3

Among his other breakthroughs, Einstein came up with a **formula** for figuring out energy (E).

A WONDROUS YEAR

Breakthroughs

In 1905, Einstein published four papers that completely changed the world of physics. This was called his *annus mirabilis*, or wondrous year:

First paper (published June 9, 1905): Einstein showed that light could be both a wave and a particle.

Second paper (published July 18, 1905): Einstein offered the first experimental proof of the size of atoms.

Third paper (published September 26, 1905): Einstein set down the mathematical theory for special relativity.

Fourth paper (published November 21, 1905): Einstein showed that the relativity theory led to the equation $E=mc^2$.

Einstein said there is no LOGICAL route to the discovery of FUNDAMENTAL LAWS. There is only the way of INTUITION, which is helped by a feeling for the ORDER that might lie behind the APPEARANCE of things.

ANNUS MIRABILIS

In 1905, while he was still working at the patent office, Einstein **revolutionized** the world of physics. He had completed his higher Doctor of Philosophy degree from the University of Zürich in April of that year, but it was not this that brought him worldwide fame. It was the four papers he published later that year in the scientific journal *Annalen der Physik* (*Records of Physics*). Einstein was just 26 years old.

Einstein...went to bed for two weeks to recover.

Einstein enjoyed his work at the patent office, even though he was turned down for a promotion!

The four papers were the results of all the thinking Einstein had done since he was very young. Each paper changed forever our views about energy, mass, space, and time. Together, they laid the foundation for modern physics. Scientists had long made suggestions about the true nature of light, but it was Einstein who confirmed how we understand light.

While other scientists were still debating whether atoms existed, Einstein not only showed that they existed but also figured out their size. Next, he turned his attention to the relationship of objects in motion. That paper, which would later become his Special Theory of Relativity, was produced in just five weeks. Einstein was so exhausted after he had finished it that he went to bed for two weeks to recover.

Ideas that changed the world

Einstein wrote a friend that his first paper would deal with radiation, the second would determine the size of atoms, the third would prove that bodies have an observable ***random*** *movement—and that the fourth was still only a rough draft.*

Exploring the ideas

The fact that, within a few months, Einstein was able to produce papers on four different subjects that completely changed physics is an **intellectual** feat that has rarely been equaled. Only Sir Isaac Newton's own *annus mirabilis* in 1666 equaled Einstein's achievement. Einstein was aware of the impact his papers would have and looked back on the period as the happiest of his life. He was so convinced that his ideas were correct that he did not feel it necessary to include experiments to prove his theories. This helped to open the era of theoretical physics and lay the foundations for the exploration of space.

Einstein's theory predicted that, if an atom split apart, it would release energy.

HISTORY'S STORY

Everyone has heard of Einstein's equation $E=mc^2$. It states that the energy (E) released from an object is equal to the mass of the object (m) times the speed of light (c), multiplied by itself. That means that if an atom is split in two, the two pieces will be smaller than the original whole because part of the mass will be released as energy. The theory forms the basis for nuclear energy.

LIGHT AND ATOMS

Einstein's four papers contained very complex ideas that are difficult to grasp, even for scientists. The first was about the photoelectric effect. The second was about Brownian motion, which makes molecules and atoms vibrate. The third paper's Special Theory of Relativity discussed the behavior of matter and energy close to the speed of light. The final paper was about how mass and energy are related to one another.

The photoelectric effect

Scientists had debated the nature of light for centuries before Einstein. By light, they meant all forms of electromagnetic radiation. This includes not just visible light but also invisible microwaves, radio waves, X-rays, and other forms of radiation. Some people believed that light traveled in waves, while Sir Isaac Newton had argued that light was a stream of particles. Einstein reached his conclusions by studying how metals release electrons when a beam of light is shone on them. Einstein's revolutionary idea was to show that light could travel as both individual particles and as a wave. He said that each of these particles carried a **quantum**, or fixed amount, of energy. The discovery that light could be both waves and particles challenged Newton's laws and laid the foundation for a new kind of physics.

This German stamp celebrates how, in 1921, Einstein won the Nobel Prize for his work on the photoelectric effect.

Brownian motion

In his second paper, Einstein figured out the size of atoms by dissolving sugar in water and figuring out the size of the sugar molecules. He showed that atoms existed by using Brownian motion to explain his reasoning.

Brownian motion is the name given to a discovery in the early 1800s by the English scientist Robert Brown (1773–1858). Brown noticed that lightweight particles of pollen carried out random movements when they floated on water. The same thing happened with dust particles. This suggested to Brown that these movements were not a sign of life but the result of some kind of force acting on the dust or the water. Einstein took the theory further, identifying that molecules in a liquid move randomly. He concluded that the atoms in the water were moving, and that atoms might receive random impacts, or bumps, of random strength from any direction.

Einstein also studied how dissolving sugar in water made the water more viscous, or thicker.

THEORIES OF RELATIVITY

Einstein is most famous for his Special Theory of Relativity, which changed not just physics but the world. He set out by figuring that light is the fastest thing in the universe, and proving that the speed of light is always constant. Einstein argued that matter and energy would behave differently if they traveled at close to the speed of light—they would not be able to obey Newton's laws. Einstein also argued that Newton's laws would not apply to objects that were very distant in space.

A comet moves through space relative to Earth—but Earth is also moving relative to the comet.

In 1887, two American scientists, Albert Michelson and Edward Morley, had set out to prove that an invisible substance, called ether, surrounded everything in the universe. They set up an experiment to prove its existence. They divided a light beam in two and bounced the two beams in different directions. They wanted to show that the two beams traveled at different speeds, proving Earth was moving through ether, which slowed the light at different rates. Every time, however, the speed of the two beams was the same. The Michelson-Morley experiment was deemed a failure. When Einstein showed that the speed of light is constant no matter what, scientists realized the Michelson-Morley experiment had not been a failure—but that no one had realized what it had proven.

Einstein saw space-time as being like a sheet. It is bent out of shape by heavy objects, like the surface of a trampoline. The heavier the object, the deeper the hole it forms.

Special Relativity

Einstein's law had two parts to it: the Theory of Special Relativity, which he came up with in 1905, and the Theory of General Relativity, which he figured out in 1915. The Theory of Special Relativity stated that all movement is relative to other objects. For example, if you look out the window when you are in a moving car, the surroundings whiz past. But if another car pulls alongside moving at an identical speed, it seems as though you are not moving at all. Although this part is fairly easy to understand, things then get very complicated: space and time are connected, so time can speed up if you are moving slowly or slow down if you are moving quickly. Together, space and time are called space-time. Special Relativity also says that mass and speed are connected, so if a ball started moving very fast, it would gain mass.

General Relativity

The Theory of General Relativity suggested a new way to think about gravity. It suggested that, if an object is a long distance away, it is impossible to tell whether any change in its motion is caused by gravity. In addition, if an object is viewed from an object that is moving because of its gravitational pull, it is impossible to tell if that distant object is moving. Going beyond that, it explained that space-time is like a trampoline that can be bent by the gravity of planets and stars.

EINSTEIN'S CONTEMPORARIES

Einstein was just one of the many physicists and mathematicians who changed the way people understood the world during the late 1800s and early 1900s. Here are some others:

Niels Bohr (1885–1962)

The Danish physicist won the Nobel Prize for Physics in 1922 for confirming the model of the hydrogen atom. A pioneer of quantum mechanics, he argued with Einstein, who did not believe in quantum theory. The two men were, however, great friends.

Edwin Hubble (1889–1953)

This American astronomer was famous for his work at Mount Wilson Observatory, California. In 1925, Hubble became the first astronomer to prove the existence of other galaxies besides the Milky Way. Hubble's Law showed that galaxies are moving away from the Milky Way at a speed in proportion to their distance from it. The Hubble Space Telescope (HST), launched into orbit around Earth in 1990, was named in his honor.

Hubble's discovery that galaxies were moving away from Earth led to a theory that the universe is expanding.

Hendrik Lorentz (1853–1928)

This Dutch physicist developed the mathematical approach used by Einstein to explore relativity. His work is known as "transformation equations."

Max Planck (1858–1947)

Born in Germany, Planck revolutionized physics with his discovery, known as Planck's constant, that energy carried by a particle has an exact value, or quantum. He was awarded the Nobel Prize for Physics in 1918. He edited the *Annalen der Physik* and was the first to read Einstein's work, becoming a father figure to the younger man.

Erwin Schrödinger (1887–1961)

Also a winner of the Nobel Prize for Physics (1933), the Austrian physicist is best known for his Schrödinger's cat "thought experiment." This tried to explain the problems of quantum mechanics: that two completely contradictory things can be true at the same time—for example, that light can be both a particle and a wave. The experiment suggests a situation in which a cat is placed in a sealed box with a hammer, a vial of poison, a source of **radioactivity**, and a Geiger counter. If the Geiger counter detects radioactivity, it releases the hammer, which will smash the vial and poison the cat. From the outside, there is no way of knowing if the cat is alive or dead. Thus, Schrödinger argued, the cat was essentially dead and alive at the same time.

According to Schrödinger's theory, it is possible—in the world of quantum mechanics—for a cat to be both alive and dead at the same time.

CHAPTER 4

Light from the major galaxy nearest to the Milky Way, the Andromeda galaxy, takes 2.5 million years to reach Earth.

EXPLAINING THE UNIVERSE

Spreading Ideas

After his *annus mirabilis* in 1905, Einstein continued to make amazing advances in physics:

1907: His Principle of Equivalence states that **acceleration** and gravity are the same thing.

1911: He publishes his prediction that the Sun's gravity would bend light from distant stars.

1915: He announces his General Theory of Relativity, explaining that gravity is the bending of space-time.

1917: He applies his General Theory of Relativity to the whole universe.

1919: The eclipse of the Sun confirms the General Theory of Relativity.

1920s: Einstein works on a theory that the universe is in a constantly changing state.

Einstein said that any INTELLIGENT FOOL could make things BIGGER and more COMPLEX, but that it took a touch of GENIUS —and a lot of courage— to move in the OPPOSITE DIRECTION.

LIGHT BENDS

In 1916, Einstein published his Theory of General Relativity, which he had been working on since 1905. **World War I** (1914–1918) had got in the way of his research. It was not until 1919 that a solar eclipse confirmed Einstein's prediction that gravity bends light in space.

...compared with this problem, the Theory of Relativity was child's play.

When an eclipse blocks the light of the Sun, the flares and other light that surround it become more easily visible.

The General Theory of Relativity came about in stages. Building on his Special Theory, Einstein realized that gravity and acceleration were very similar. He figured out that the gravity of a huge object, such as the Sun, would make light bend as it passed. Until he had the math, however, Einstein struggled to explain his theory. He grumbled that, compared with this problem, the Theory of Relativity was child's play.

The chance to prove the theory came in 1919, during the total solar eclipse. During a solar eclipse, the moon blocks sunlight reaching Earth. This creates darkness that makes it possible to photograph the stars beyond the Sun, which are usually hidden by sunlight. Einstein said that such photos would show light bending as it passed other planets and the Sun. He was correct.

Ideas that changed the world

Einstein said that when a blind beetle crawled over the surface of a curved branch, it did not notice that the track it covered was curved so it kept crawling in a circle. He was lucky enough to notice what the beetle did not notice.

Exploring the ideas

Einstein's genius was to think of the universe in a way that nobody else had thought of it. He thought that space was like a vast sheet, and that large objects caused space to bend. The amount of the bend depended on the mass of the object, the same as when you place heavier or lighter balls on a sheet: the heavier balls cause deeper hollows. Einstein concluded that space was not a fixed thing but could change constantly. It was shaped by time as well as its other dimensions, such as length and width. That was a revolutionary idea.

Einstein used the example of a beetle to express his idea that he just happened to notice small details that revealed a bigger picture of the universe.

HISTORY'S STORY

To test Einstein's theory that light bent as it traveled through space, all that was needed was a total **solar eclipse**. Einstein's friend, the British astronomer Arthur Eddington, arranged for teams to photograph the 1919 eclipse from the west coast of Africa and in Brazil. The photos showed the light of distant stars bending. This proved Einstein's theory was correct.

FAMOUS MAN

The 1919 eclipse had proven Einstein's General Theory of Relativity to be correct. Few people truly understood what the theory meant, but suddenly Einstein became famous. With his wild hair and odd clothing, Einstein was everybody's idea of a mad professor. He was shocked by fame. Everybody wanted to talk to him; lecture requests flooded in from around the world. There was no time to focus on his research.

In 1913, Einstein quit the patent office and had taken a job as director of the Kaiser Wilhelm Institute for Physics in Berlin, Germany. Now people flocked to his lectures in such large numbers that it was hard for his actual students to get in. His book, *Relativity: The Special and General Theory*, had been published in German during World War I. Now it was a bestseller translated into English and French.

In 1921, President Warren G. Harding (center) invited Einstein (on his right) to the White House.

Einstein's fame did have some benefits. He realized that people would listen to him, so he could draw attention to causes he believed in. These included **pacifism**, or opposition to war, which had been important to him since World War I. Another passion was Zionism, the movement to establish a homeland for the Jews. The hostility toward Jews in Germany led him to tour the United States in 1921 to raise money to build a Hebrew University in Jerusalem, which was then under British control.

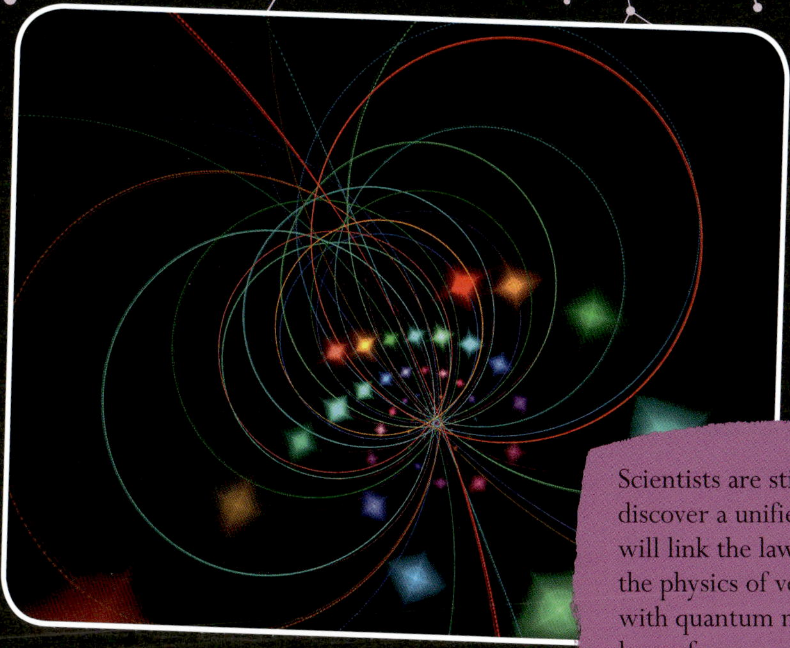

Scientists are still trying to discover a unified theory that will link the laws that cover the physics of very big objects with quantum mechanics, the laws of very small objects.

The Nobel Prize for Physics

Albert Einstein was so famous for his Theory of Relativity that everyone assumed he would be next in line for the Nobel Prize. However, a group of anti-Jewish German scientists tried to argue that relativity was a Jewish conspiracy. They campaigned to prevent Einstein from being given the award in 1920. The following year, the Nobel committee did not award a prize for physics. In 1922, the Danish physicist Niels Bohr was awarded the prize for his work on atoms, but it was also announced that Einstein had won the unawarded 1921 prize for his work on the photoelectric effect.

A unified field theory

During the 1920s, Einstein set out to link gravity and quantum mechanics, or the study of the inner workings of atoms. A theory that links these two fields is known as a **unified field theory**. Einstein believed it was possible to do this through a set of equations. He published his first paper on the subject in 1929, even though he was aware that his theory had faults. He kept working on his unified field theory for the rest of his life, with no success.

INTO EXILE

Einstein had watched the **Nazi Party** and its leader, Adolf Hitler, gain popularity in Germany during the 1920s. Hitler hated Jews. In 1933, after the Nazis achieved success in national elections, Hitler became the **Chancellor** of Germany and began putting laws in place to persecute Jews. Einstein realized he and his family would have to leave Berlin.

Adolf Hitler encouraged Germans to blame Jews for all Germany's problems.

Einstein had spent a lot of time in the 1920s outside Germany, giving lectures and touring. When he was in Berlin, however, it was impossible to ignore the rise of the Nazis and their hostile policies toward Jews. The Nazis hated Jews, pacifists, and intellectuals. Not only was Einstein all three—he was also one of the most famous citizens of Germany. This made him a target for the Nazis.

When Hitler came to power in January 1933, Einstein and his second wife, Elsa, were in the United States. The Nazis raided Einstein's Berlin home and seized his sailboat, which they later sold. They threatened that Einstein was a target for assassination, or being killed, in Germany, even though he was the toast of the rest of the world. It was clear that Einstein's old life in Germany was over. He never returned to the country of his birth.

A new life in America

When Einstein got back to Europe, he stayed in Brussels, in Belgium. Although he was a Swiss citizen, he had been given a German passport after World War I. Now he marched to the German consulate and handed his passport back.

Einstein was not sure what his next move should be. He saw that Hitler's plans to expand Germany made it unwise to stay in Belgium. Since the publication of his Theory of Relativity, there was no shortage of offers of teaching jobs. Oxford University offered him a teaching job, but Elsa did not want to settle in England. Finally, he was offered a position at the Institute for Advanced Study in Princeton, New Jersey.

In October 1933, the Einsteins set sail for New York. Einstein's oldest son, Hans Albert, moved to America in 1938, followed by Einstein's sister Maja in 1939. His younger son, Eduard, remained safely in Switzerland, which was not invaded by Germany. They never saw each other again.

Einstein went into exile with his beloved second wife, Elsa, whom he had married in 1919.

HISTORY'S STORY

Many scientists fled Germany when the Nazis took control in 1933, especially Jews like Einstein. Between 1939 and 1945, the Nazis **systematically** murdered 6 million Jews during the Holocaust. Among the dead were many Jewish scientists and thinkers, including the chemist Ernst Cohen and the mathematician Juliusz Schauder.

EARLY TWENTIETH-CENTURY SCIENTIFIC BREAKTHROUGHS

1900 Max Planck sets out his Quantum **Hypothesis**, which argues that light energy can only travel in bundles called quanta.

1905 Albert Einstein's four papers prove that light is both a wave and a particle and that atoms exist. He also outlines the Theory of Special Relativity.

1913 Niels Bohr sets out a model for the structure of the atom.

1916 Einstein publishes his Theory of General Relativity.

1926 Erwin Schrödinger's equation describes the nature of wave particles.

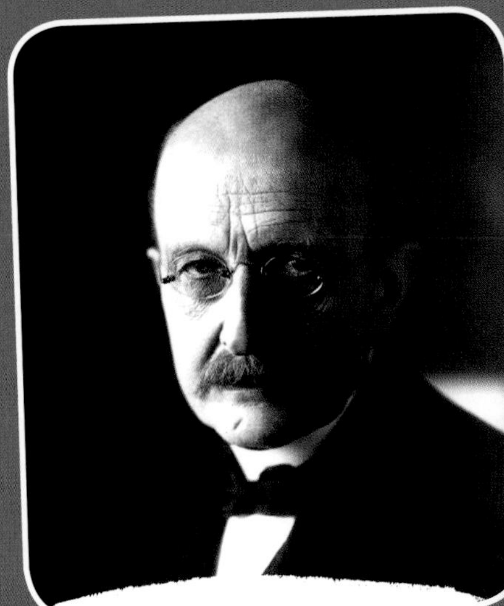

As an established physicist, Max Planck popularized Einstein's theories and helped the younger man make a name in the scientific world.

1927 Werner Heisenberg produces his uncertainty principle, which says it is possible to know either the location or the speed of a subatomic particle, but not both at the same time.

1929 Edwin Hubble confirms that the universe is expanding.

1938 Otto Hahn carries out the first nuclear fission, splitting atoms and paving the way for the atom bomb.

1948 Richard Feynman publishes his quantum electrodynamics theory, proposing a new way to understand the nature of waves and particles.

U.S. bombers dropped atomic bombs on Japan in August 1945, at the end of a long period of learning about the nature of atoms.

CHAPTER 5

This photograph shows Einstein in 1947. His appearance is what many people think of when they imagine a scientist.

NATIONAL FIGURE

Reputation

Here are some of the honors that Einstein received, before and after his death:

1921: Nobel Prize for Physics by the Nobel Institute, Sweden

1925: Copley Medal by the Royal Society, England

1929: Max Planck Medal by the German Physical Society

1931: Prix Jules Janssen by the French Astronomical Society

1935: Franklin Medal by the Franklin Institute in Philadelphia, United States

1999: Person of the Century by *Time* magazine, United States

2005: World Year of Physics created in honor of the 100th anniversary of Einstein's *annus mirabilis*

Einstein said there were only two ways to live one's life. One is as though NOTHING is a MIRACLE. The other is as though EVERYTHING is a MIRACLE.

ATOMIC BOMB

Einstein's equation $E=mc^2$ had paved the way for scientists to split atoms, the smallest particles that can exist independently. In 1938, German scientists in Berlin discovered a way of splitting atoms of uranium, which Einstein had thought was impossible. In theory, it was possible to start a **chain reaction** in which splitting atoms caused more atoms to split. This would release huge amounts of energy in an explosion—if anyone could figure out how to use the technology to build a bomb.

The atomic bomb was just a theory, but the clock was ticking. Einstein wrote to President Franklin D. Roosevelt to warn him that the Germans were secretly working on a bomb that would unleash the power contained within atoms. Einstein's advice led to the top-secret **Manhattan Project**, which resulted in the end of **World War II** when two atomic bombs were dropped on Japan in 1945.

Einstein never worked on the Manhattan Project. The U.S. authorities were suspicious of him, because of his German birth and his involvement with peace groups. Many of his fellow scientists joined the project, however. Einstein was worried that no one understood what they were building. He wrote Niels Bohr, saying that politicians did not understand the full threat an atomic bomb would pose.

The Manhattan Project was based at Los Alamos in New Mexico.

Ideas that changed the world

Einstein said concern for humans and their fate must always be the main interest of science. He warned other scientists not to forget this in the midst of their diagrams and equations.

Exploring the ideas

Einstein was a pacifist and the thought of his famous equation being the starting point for the creation of a deadly weapon filled him with horror. As the Manhattan Project continued, Einstein wrote a second time to Roosevelt, asking the president to ensure that the atomic bomb was not dropped on Japan. President Roosevelt died on April 12, 1945, however. Einstein's letter was found, unopened, in his office. When the first atomic bomb was dropped on Hiroshima on August 6, 1945, more than 80,000 people were killed. Another 70,000 died when a second bomb was dropped on Nagasaki three days later. The bombs led to Japan's surrender and the end of World War II (1939–1945).

This ruined building was one of the few structures that survived the blast of the bomb at Hiroshima.

HISTORY'S STORY

Einstein was not the only scientist to fear atomic weapons. Einstein's friend Leo Szilard (1898–1964) had figured out the circumstances needed to create a nuclear chain reaction. He appreciated the dangers presented by such a weapon. So did the physicist Eugene Wigner (1902–1995), who warned that anyone with uranium could build a devastating bomb.

STANDING UP

In 1940, Einstein became an American citizen. Living in the United States, he continued to be an outspoken pacifist and anti-racist. He was almost as famous for his non-scientific opinions as he was for his physics. With his crazy hair and his mismatched clothes, he was a distinctive figure who was often in the public eye.

Throughout his life, Einstein was a pacifist. He had argued against World War I. Yet, as a Jew who opposed the Nazis, he felt torn during World War II. Although his advice to Roosevelt helped speed up American development of the atomic bomb, he later tried to caution the president against using it. When news broke of the bombing of Hiroshima, Einstein apparently said in German, "Oh, woe."

Albert Einstein is congratulated after the ceremony that made him an American citizen in 1940.

Einstein continued to argue against nuclear weapons ...

After the war, scientists formed the Emergency Committee of Atomic Scientists. They believed the world needed an agreement about weapons control that would prevent any more atomic bombs being dropped. In May 1946, Einstein was named as their first chairman. Einstein continued to argue against nuclear weapons, declaring that if he had known that German scientists would not succeed in producing an atomic bomb, he would never have suggested it to the Americans.

Civil rights

When he first moved to the United States, Einstein was appalled by the system of **segregation** used to separate races. He felt it was his duty to speak out against racism and often did. He made a point of mixing with African Americans in Princeton. At the time, Princeton University, like other schools, barred African Americans from attending.

In 1937, when the opera singer Marian Anderson was not allowed a hotel room in Princeton because of her race, Einstein invited her to stay at his home. The two became good friends, but his action shocked many Americans.

In 1939, when Marian Anderson was banned from singing in a Washington, DC, hall because she was black, she sang to 75,000 people from the steps of the Lincoln Memorial.

Zionism

Einstein was not a practicing Jew, but he believed strongly in the establishment of a Jewish homeland. His first visit to the United States in 1921 had been to raise money for the World Zionist Organization. After the horror of the Holocaust, the cause became even more important to him. He did not agree with the new state of Israel being located in Palestine, but his views were ignored. After the state of Israel was created in 1948, Einstein declared it was "the fulfillment of our [Jewish] dreams." He was even offered the ceremonial position of president. He turned the offer down, saying that he had "neither the natural ability nor the experience to deal with human beings."

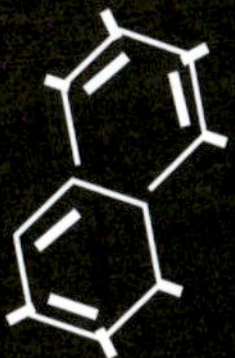

FINAL DAYS

The last years of Einstein's life, before he died in 1955, were part of the long period of tension between the United States and the **Soviet Union** called the Cold War. Einstein had created the equation that made the atomic bomb possible. Now, as tensions grew, the big worry was whether the Soviets would be able to build their own bomb.

The United States was gripped by a fear of **communism**, which was the system of government in the Soviet Union. During the 1940s and 1950s, Wisconsin senator Joseph McCarthy carried out a witch hunt for suspected communists in every corner of American life. Einstein was appalled. He believed in intellectual and personal freedom and supported the right of individuals to refuse to appear in front of McCarthy's House Un-American Activities Committee. Among those accused of being communist was J. Robert Oppenheimer, the man who had developed the atomic bomb. Einstein's support helped Oppenheimer keep his job. Oppenheimer later said Einstein was the nicest man he ever knew.

Joseph McCarthy believed that Hollywood was full of communists.

Although Einstein devoted himself to many political causes, as well as giving lectures and teaching at Princeton, he also tried to work on his theories. He was still working on his unified field theory, which he had begun in the mid-1920s, but he never completed it. By 1948, he was in very poor health. He was also lonely. His beloved second wife, Elsa, had died in 1936. His sister Maja died in 1951.

The end

Einstein's last few years were spent mostly in and around his house in Princeton. On April 18, 1955, Einstein died in hospital following a short illness. Just a few days earlier, he had been working on equations and writing letters. The last letter he wrote urged all nations to give up nuclear weapons.

Einstein's brain

Shortly after his death, an **autopsy** was performed on Einstein's body to discover the cause of his death. Unknown to Einstein's family, the physician who performed it, named Thomas Harvey, removed Einstein's brain. Harvey said he wanted to study it to understand genius, but he had no background in brain research. The family tried to get the brain back, but Harvey kept it. However, he did send samples to scientists who asked for them. Proper research on Einstein's brain was carried out only in the late 1990s. Research showed that Einstein's parietal lobe, the part of the brain responsible for mathematical and spatial reasoning, was 15 percent larger than in average brains.

Einstein lived in this house on Mercer Street in Princeton from 1935 until his death.

EINSTEIN'S KEY WORKS

Special Theory of Relativity **(1905)**

Einstein's breakthrough work, discussed the relationship between space and time.

The Foundation of the General Theory of Relativity **(1916)**

The result of 10 years' work, this book included the mathematical formula needed to show that gravity bends light.

This image from the HST shows light being bent by the gravity of a distant star, creating a lens-like effect.

Investigations on the Theory of Brownian Movement **(1926)**

Einstein gave proof that atoms exist, building on the paper he published in 1905.

About Zionism: Speeches and Letters **(1931)**

This was a collection of Einstein's speeches and lectures on Zionism.

Why War? (1933)

This pamphlet contained the letters Einstein and the **psychoanalyst** Sigmund Freud (1856–1939) wrote to each other about their views on human nature and how to stop violence and war.

Essays in Science (1934)

Einstein presented 16 essays that explained the work of other scientists and physics in general, aimed at people who had no background in science.

Out of My Later Years (1950)

Another collection of essays, this presented Einstein's views on topics that fascinated him in the worlds of science, politics, and history.

Sigmund Freud believed the mind has a powerful subconscious that controls much of people's lives, even if they are not aware of it.

CHAPTER 6

This is an artist's impression of a black hole. The first, blurred photo of a black hole was not taken until 2019.

WORLD OF SCIENCE
Legacy

Einstein's theories have led to many important breakthroughs:

1967: The Theory of Weak Interaction explains how subatomic particles create radiation.

1972: Stephen Hawking and others put forward new laws about black holes.

1974: The charm **quark**, an **elementary** particle, is discovered.

1977: The bottom quark is discovered.

1981: The Theory of Cosmic Inflation explains the expansion of space in the moments after the Big Bang.

1998: The accelerating universe is discovered: the speed at which galaxies are moving away from us is increasing.

2012: The Higgs boson elementary particle is discovered.

2015: Gravitational waves, disturbances in space-time, are detected.

Einstein said that IMAGINATION is more important than KNOWLEDGE. While knowledge is LIMITED, imagination encircles the WORLD.

AN EINSTEIN UNIVERSE

Since Einstein's death, scientists have advanced his unified field theory. They have made new discoveries, such as the mysterious strong force, which were missing from his calculations. In addition, space telescopes and radio telescopes continue to observe new phenomena in deep space that confirm many of his theories. Einstein's reputation has continued to grow.

Einstein's legacy is unmatched. Scientists continue to discover new facts he accurately predicted at a time when the technology did not yet exist to confirm his ideas. However, he still made sure his science was based on proof by using math and observation, just as Galileo and Newton had used experiments to prove their points.

The HST orbits above the distorting effects of Earth's atmosphere.

When Einstein was alive, black holes and the Big Bang were just theories. Today, they are the basis of modern **cosmology**. Einstein's work allowed scientists to unlock the universe's secrets. When the National Aeronautics and Space Administration (NASA) launched the Gravity **Probe** A in 1976, it proved his theory of how gravity affects space-time. The probe carried a clock which, as Einstein had suggested, showed that time passed slower on its journey through space than on an identical clock on Earth. In 2004, Gravity Probe B measured the bending of space-time around Earth. It again confirmed the General Theory of Relativity.

Einstein said that humans still do not know one thousandth of 1 percent of what nature has revealed to us.

Exploring the ideas

Einstein's theories of relativity opened up the universe for future generations. He also opened up a new branch of physics: theoretical physics. Some of the technology we take for granted exists thanks to Einstein's ideas, including TVs, digital cameras, lasers, and the global positioning system (GPS). Even the paper towel is down to Einstein. One of his earliest research papers, published in 1901, looked into wicking, the process by which towels soak up water even though gravity wants to drag the fluid down. His findings helped lead to the scientific development of the first paper towel in 1907.

Using signals from GPS satellites to find your location on Earth was an idea first predicted by Einstein's theories.

HISTORY'S STORY

Today's stock markets see billions of dollars traded every day. Einstein's paper on Brownian motion, published in 1905, included a formula to express the apparent randomness of the action that moves atoms. It turns out it can also be applied to the ups and downs of the stock market. It still serves as the basis for some stock market forecasts.

THE QUANTUM WORLD

Since Einstein's death, there have been many advances in quantum mechanics. The discovery that the universe is made up of particles that are even smaller than atoms meant that ordinary physics was no longer helpful. A new set of laws had to be invented. Before Einstein, scientists had tried to prove that the universe ran according to fixed laws. After Einstein, it became clear it did not.

This illustration shows an atom breaking apart into many **elementary particles**.

In 1926, the German physicist Werner Heisenberg formulated his famous uncertainty principle. He said that to be able to predict the future position and speed of a particle, you had to be able to precisely measure its current position and speed—but that this was impossible. The more accurately you measure the speed, the less accurately you can measure its position, and vice versa. It is possible to know the particle's speed or its position, but not both. Heisenberg's uncertainty principle has become the basis of all quantum mechanics.

Quantum mechanics cannot predict a single outcome to a problem. Instead, it predicts lots of different outcomes, then figures out which is the most likely. As scientists have found out more about the particles that make up the universe, quantum mechanics has become more important.

What's the universe made of?

Until the 1960s, scientists thought that protons and neutrons were the smallest particles on which everything else was built. Experiments in particle accelerators have started to prove that there are even smaller subatomic particles, such as quarks and gluons. Particle accelerators are machines that send particles to super-high speeds.

Physicists are now using incredibly complex experiments to identify the most basic building block of the universe. What is the elementary particle from which everything else is made? Or is there a smaller particle yet to be discovered? Trying to solve this question meant building the world's largest and most complex scientific instrument, the Large Hadron Collider (LHC), underground in Switzerland. The machine is a long tube in which subatomic particles are forced to collide with each other at close to the speed of light, re-creating what happened in the universe moments after the Big Bang. In 2012, the LHC observed a new elementary particle. This may be the Higgs boson, which some people believe is the smallest particle of all.

The LHC uses powerful magnets to accelerate atoms to close to the speed of light.

The LHC is based in a circular tunnel 16.8 miles (27 km) around.

A MODEL SCIENTIST

Einstein is the most famous scientist of the 20th century. This is not just because of his contributions to science, which most people do not understand. It is because much of his life was devoted to helping the causes he believed in. His lack of interest in the clothes he wore and the food he ate made him seem like everyone's idea of an eccentric professor.

Einstein gave hundreds of lectures around the world on scientific and political subjects.

Lurking behind Einstein's crazy hair and unusual way of dressing was a model for modern physicists. Einstein was absent minded but also incredibly intelligent. He could consider the vastness of space, yet be sensitive to other people's needs. He came up with the formula that allowed the atom to be split but was terrified by its consequences for humanity. He spent a lot of his time after World War II arguing against atomic weapons.

Einstein was a very public figure. He wrote to newspapers giving his views, and he gave lectures not just to the scientific community but also to ordinary people. He loved getting letters from children, asking him about physics. He wrote back to them in simple language they could understand. Einstein was an expert communicator long before the days of social media.

A new science

Einstein made science popular. He also made it seem fun. Many people became scientists because Einstein showed them that there were all kinds of areas that they could work in. He showed that science must be based on proof but that coming up with hypotheses, or possible explanations, was a vital part of a scientist's work. Without imagination and leaps of faith, he argued, new breakthroughs could not be achieved. But once an idea had been suggested, it was the scientist's responsibility to test and test until their idea was proven.

For Einstein, science had to be part of life along with everything that makes a person human—a love of culture, thinking, and just living life. He inspired artists and writers who wanted to describe the world he was telling us about.

Einstein was a skilled violinist. When he was younger, he played in a group with physicist Max Planck.

HISTORY'S STORY

Einstein loved music and was a talented violinist. He played the violin his whole life, taking it with him wherever he went. He once said that the most joy he had in his life came from his violin. He believed that music and science were closely aligned. Playing the violin helped him think. He would often play the violin at fundraisers for his political causes.

MODERN-DAY SCIENTISTS

Entire generations of scientists based their careers on the breakthroughs made by Einstein. Here are four:

Sir Roger Penrose (1931–)

An English mathematical physicist, Penrose worked on general relativity and cosmology. He revolutionized the mathematical tools used to analyze the properties of space-time. He worked with Stephen Hawking and calculated many of the basic features of black holes.

Stephen Hawking spent much of his time trying to understand black holes. They bend space-time so much that not even light can escape.

Stephen Hawking (1942–2018)

Probably the most famous scientist in the world since Einstein, Hawking was an English theoretical physicist whose theory of exploding black holes drew from relativity theory and quantum mechanics. He suffered from motor neurone disease, which eventually made him wheelchair bound and unable to speak without a computer-aided voicebox. Despite this, Hawking lectured and toured the world. His *A Brief History of Time* (1988) sold more than 10 million copies.

This HST image of a distant part of the universe shows clouds of gas where billions of stars are being formed.

Sir Fred Hoyle (1915–2001)

This British mathematician and astronomer is best known as the biggest defender of the steady-state theory of the universe, which is now rejected by most cosmologists. This theory states that the universe is expanding and that matter is being continually created to keep the **density** of matter and space constant.

Richard Feynman (1918–1988)

An American theoretical physicist, Feynman was widely seen as the most brilliant and influential physicist of his generation. He reinvented quantum electrodynamics and was awarded the Nobel Prize for Physics in 1965.

GLOSSARY

acceleration Increase in speed
astronomers Scientists who study space
astronomy Study of the stars, planets, and space
autopsy Examination of a dead body to figure out the cause of death
Big Bang The moment the universe began
black holes Regions of space-time from which nothing can escape because their gravity is so strong
chain reaction A series of events, each caused by the previous one
chancellor Head of government
chemistry The branch of science that studies the characteristics of substances
communism A form of government in which everything is shared equally between people
density A measure of how tightly packed molecules are in something
electrodynamics The branch of science that studies the relationship between electric currents and magnetic fields
electrons Subatomic particles with a negative electric charge that circle the nucleus of an atom
elementary particles Subatomic particles that are not made of other particles
equation A mathematical statement that says two things are equal
formula A mathematical rule written with symbols
formulated Expressed in a precise form
galaxy A system of millions or billions of stars, gas, and dust held together by gravity
geology The study of rocks and Earth
gravity The force that pulls all objects toward each other
hypothesis A proposed explanation for something, based on scientific observation
intellectual Involving a person's abilities to think and to understand ideas and information
Manhattan Project Code name for an American project to develop an atom bomb
mass A measure of the amount of matter in an object
matter Physical substance
molecules Groups of atoms bonded together
Nazi Party A German political party (1920-1945) that tried to wipe out Jewish people and others it considered to be "inferior"
neutrons Subatomic particles without an electric charge that are in the nucleus of most atoms
Nobel Prize An international prize awarded for outstanding achievement
pacifism The belief that war is wrong
particles Tiny pieces of matter
patent A government license recognizing an invention belongs to the inventor
philosophy The study of the nature of reality
physicist A scientist who studies the nature and properties of matter and energy
physics The branch of science concerned with the nature and properties of matter and energy
probe An uncrewed spacecraft
protons Positively charged subatomic particles that are in the nucleus of most atoms
psychoanalyst Someone who treats people with mental health problems
quantum The smallest unit of energy
quantum mechanics The branch of science that deals with the behavior of subatomic particles
quark An elementary particle; protons and neutrons are each made up of three quarks
radioactivity The breakdown of one type of atomic nucleus into another
random Lacking a definite plan, purpose, or pattern
segregation Enforced separation of different races
Soviet Union A group of communist countries in northern Eurasia between 1922 and 1991
systematically Done according to a set plan
theoretical physics The branch of physics that uses math to explain things that cannot be seen
theory A scientific principle suggested to explain facts and observations
Theory of Relativity, General Einstein's theory that explains how gravity can bend space-time
Theory of Relativity, Special Einstein's theory that explains how space and time are connected
unified field theory An attempt to use a single theory to describe all the universe's forces and the relationships between elementary particles
wave A disturbance that transfers energy through matter or space
World War I (1914–1918) A multi-country war in which Germany, Austria-Hungary, and their allies were defeated
World War II (1939–1945) A multi-country war in which Germany and Japan were defeated

FOR MORE INFORMATION

BOOKS

Anderson, Jennifer Joline. *Albert Einstein: Revolutionary Physicist* (Great Minds of Science). Minneapolis, MN: Core Library, 2014.

Ferguson Delano, Marfé. *Genius: A Photobiography of Albert Einstein*. Washington, DC: National Geographic Children's Books, 2015.

May, Andrew. *Albert Einstein: Scientist* (History Makers). New York, NY: Cavendish Square, 2017.

Tolish, Alexander. *Gravity Explained* (The Mysteries of Space). Berkeley Heights, NJ: Enslow Publishers, 2018.

WEBSITES

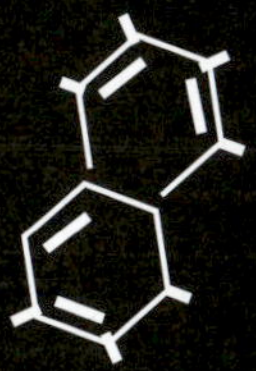

Biography—www.ducksters.com/biography/albert_einstein.php
The story of Einstein's life and times.

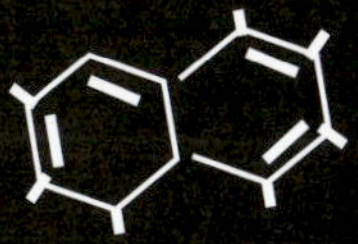

Black holes—www.nasa.gov/audience/forstudents/k-4/stories/nasa-knows/what-is-a-black-hole-k4.html
A page from NASA about black holes, which Einstein predicted.

Fun facts—www.coolkidfacts.com/albert-einstein-for-kids
A page with many interesting facts about Einstein and his work.

Relativity—www.space.com/17661-theory-general-relativity.html
An easy-to-understand explanation of Einstein's Theory of General Relativity.

Publisher's note to educators and parents: Our editors have carefully reviewed these websites to ensure that they are suitable for students. Many websites change frequently, however, and we cannot guarantee that a site's future contents will continue to meet standards of quality and educational value. Students should be closely supervised whenever they access the Internet.

INDEX